ARE YOU a MatH GENiUS?

 = 50

 = 40

 = 20

= 10

 = 100

= 30

THE INVENTOR'S
BOOK OF CALCULATION GAMES
FOR BRILLIANT THINKERS
BY SaraH Janisse Brown ~ tHE THinking TREE PUBLISHing, LLC

THE Thinking TREE

PUBLISHING COMPANY

Sarah Janisse Brown

ART LOGIC SCIENCE SPELLING READING COLOR THINKING DRAWING CREATING

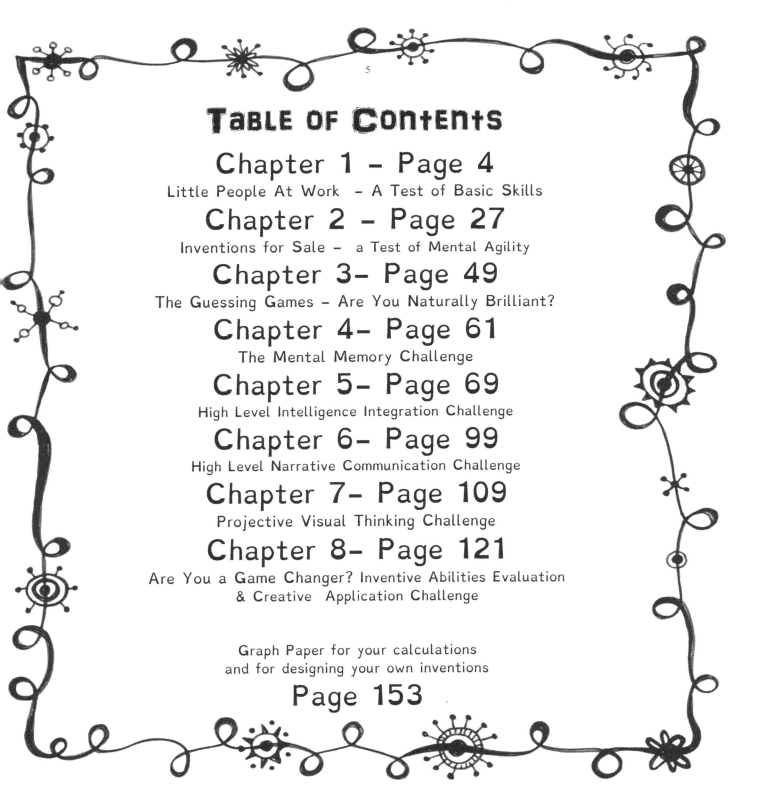

TABLE OF CONTENTS

Chapter 1 – Page 4

CHAPTER ONE

LittLE PEOPLE at WORK
A Test of Basic Skills

If you are able to answer all the questions in **Chapter One** correctly you can move on to Chapter Two, Three, Four, Five, Six, Seven and Eight! You are pretty smart, but how are you when it comes to mixing your mathematical knowledge with creativity, and logic?

The questions in this chapter are not simple. If you can answer them well you can move on knowing that you are able to use basic math skills to solve imaginary math problems.

Check the answer key in the back after writing down your answers. If you fail you need brush up on your basic addition and multiplication skills... or use a calculator. Whatever works for you is fine with us. We just want you to be a happy and clever individual.

Invention # 1
This is a Fringinwinger

How smart are you?
Look at the INVENTION on the Left.

Little employees are needed to run the
machine. Without them it can not work.

1. How many people do you see?_____
2. Each person earns $5.00 per hour.
3. How much money will you need to pay them if
   you want to operate this machine for 3 hours?

Answer:_____

Bonus Question:
How much do you think it will cost to replace
both marshmallows and hire a part time worker
to shine the boot?

Invention #2
This is a Lockinburster.

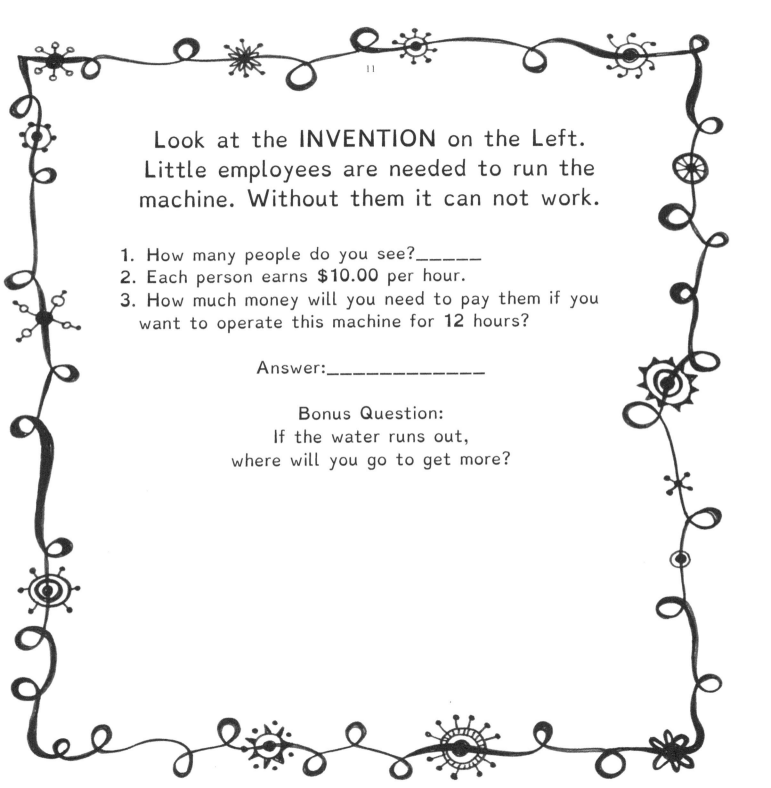

Look at the **INVENTION** on the Left.
Little employees are needed to run the
machine. Without them it can not work.

1. How many people do you see?_____
2. Each person earns **$10.00** per hour.
3. How much money will you need to pay them if you
   want to operate this machine for **12** hours?

Answer:_____

Bonus Question:
If the water runs out,
where will you go to get more?

Invention #3
This is a Fropinfosser

Look at the **INVENTION** on the Left.
Little employees are needed to run the
machine. Without them it can not work.

1. How many people do you see?_____
2. Each person earns **$7.00** per hour.
3. How much money will you need to pay them if you
   want to operate this machine for 8 hours?

Answer:_____

Bonus Question:
If someone took the stool away
what would you put in its place?

Invention #4
This is a Lunkinturkey

Look at the **INVENTION** on the Left. Little employees are needed to run the machine. Without them it can not work.

1. How many people do you see?_____
2. Each person earns **$3.00** per hour.
3. How much money will you need to pay them if you want to operate this machine for **40** hours?

Answer:_____

Bonus Question:
If you wanted to build two of these
where would you go to get a **2nd** broom?

Invention #5
This is a Stopitsprungy

Look at the **INVENTION** on the Left. Little employees are needed to run the machine. Without them it can not work.

1. How many people do you see?_____
2. Each person earns **$25.00** per hour.
3. How much money will you need to pay them if you want to operate this machine for 8 hours?

Answer:_____

Bonus Question:
How many hours should
the bugs stay in the jar?

Invention #6
This is a Decitonimeter

Look at the **INVENTION** on the Left.
Little employees are needed to run the
machine. Without them it can not work.

1. How many people do you see?_____
2. Each person earns **$2.50** per hour.
3. How much money will you need to pay them if you
   want to operate this machine for **10** hours?

Answer:_____

Bonus Question:
Who will eat the marshmallow
and hotdog when they are
finished cooking?

Invention #7
This is a Doodiepopper

Look at the **INVENTION** on the Left.
Little employees are needed to run the
machine. Without them it can not work.

1. How many people do you see?_____
2. Each person earns **$7.50** per hour.
3. How much money will you need to pay them if you
   want to operate this machine for **100** hours?

Answer:_____

Bonus Question:
If all your little employees asked
for a raise what would you do?

Invention #8
This is a Rupkinasowind

Look at the **INVENTION** on the Left.
Little employees are needed to run the
machine. Without them it can not work.

1. How many people do you see?_____
2. Each person earns **$100** per hour.
3. How much money will you need to pay them if you
   want to operate this machine for **14** days, non-stop?

Answer:_____

Bonus Question: If a bird and a rabbit came near
your machine what would you do?

Invention #9
This is a Bellihootie

Look at the **INVENTION** on the Left.
Little employees are needed to run the
machine. Without them it can not work.

1. How many people do you see?_____
2. Each person earns **$32.50** per hour.
3. How much money will you need to pay them if you
   want to operate this machine for **30** minutes?

Answer:_____

Bonus Question:
What do you need to worry about the most,
cats, dogs or rabbits?

Invention #10
This is a Slockomoggy

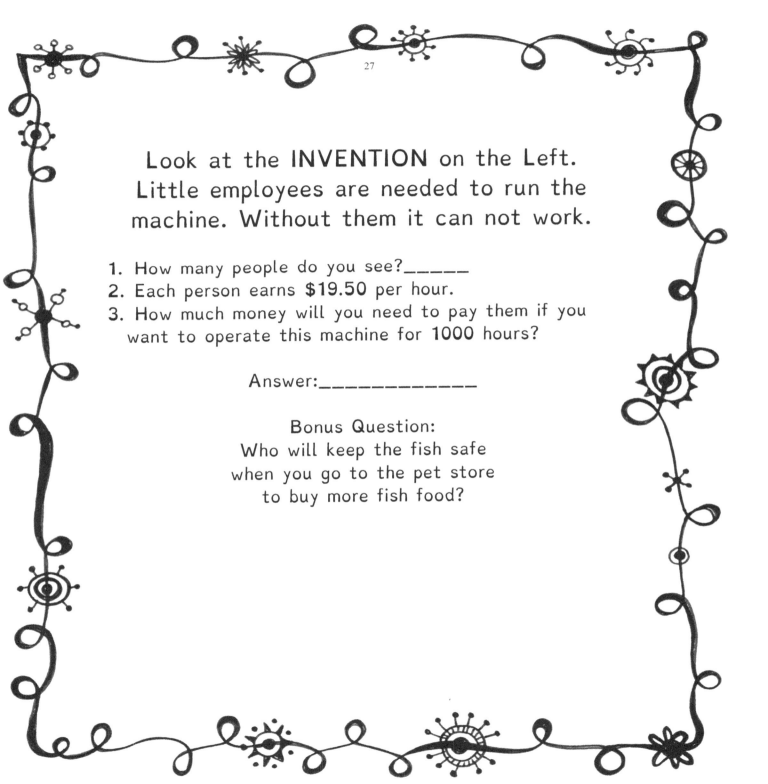

Look at the **INVENTION** on the Left.
Little employees are needed to run the
machine. Without them it can not work.

1. How many people do you see?_____
2. Each person earns **$19.50** per hour.
3. How much money will you need to pay them if you
   want to operate this machine for **1000** hours?

Answer:_____

Bonus Question:
Who will keep the fish safe
when you go to the pet store
to buy more fish food?

Chapter TWO

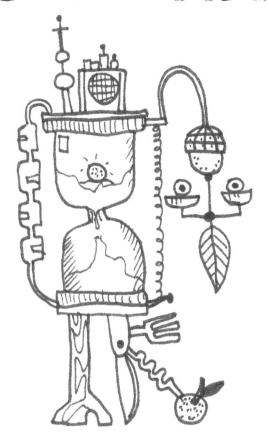

Inventions For Sale

Practice demonstrating your mathematical skills, mental agility and creativity!

Invention #1
This is a Fringinwinger

Look at the **INVENTION** on the Left.
It's for Sale.
How much does it cost?
$_____
Each of the parts below has a value.
Calculate the value of the items used to
create the invention to find the cost.

$100

$15

$25

$1

$1

$30

$30

$15

$10

$50

$20

$10

$5

$1

$10

Invention #2
This is a Lockinburster.

Look at the INVENTION on the Left.
It's for Sale.
How much does it cost?
$_____
Each of these parts below has a value.
Calculate the value of the items used to
create the invention to find the cost.

$100

$15

$25

$1

$1

$30

$30

$15

$10

$50

$20

$10

$5

$1

$10

Invention #3
This is a Fropinfosser

Look at the INVENTION on the Left.
It's for Sale.
How much does it cost?
$_____

Each of these parts below has a value.
Calculate the value of the items used to
create the invention to find the cost.

$100

$15

$25

$1

$1

$30

$30

$15

$10

$50

$20

$10

$5

$1

$10

Invention #4
This is a Lunkinturkey

Look at the INVENTION on the Left.
It's for Sale.
How much does it cost?
$_____
Each of these parts below has a value.
Calculate the value of the items used to
create the invention to find the cost.

$100

$15

$25

$1

$1

$30

$30

$15

$10

$50

$20

$10

$5

$1

$10

Invention #5
This is a Stopitsprungy

Look at the **INVENTION** on the Left.
It's for Sale.
How much does it cost?
$_____

Each of these parts below has a value.
Calculate the value of the items used to
create the invention to find the cost.

$100

$15

$25

$1

$1

$30

$30

$15

$10

$50

$20

$10

$5

$1

$10

Invention #6
This is a Decitonimeter

Look at the INVENTION on the Left.
It's for Sale.
How much does it cost?
$_____

Each of these parts below has a value.
Calculate the value of the items used to
create the invention to find the cost.

$100

$15

$25

$1

$1

$30

$30

$15

$10

$50

$20

$10

$5

$1

$10

Invention #7
This is a Doodiepopper

Look at the INVENTION on the Left.
It's for Sale.
How much does it cost?
$_____
Each of these parts below has a value.
Calculate the value of the items used to
create the invention to find the cost.

$100

$15

$25

$1

$1

$30

$30

$15

$10

$50

$20

$10

$5

$1

$10

Invention #8
This is a Rupkinasowind

Look at the INVENTION on the Left.
It's for Sale.
How much does it cost?
$_____

Each of these parts below has a value.
Calculate the value of the items used to
create the invention to find the cost.

$100

$15

$25

$1

$1

$30

$30

$15

$10

$50

$20

$10

$5

$1

$10

Invention #9
This is a Bellihootie

Look at the INVENTION on the Left.
It's for Sale.
How much does it cost?
$_____
Each of these parts below has a value.
Calculate the value of the items used to
create the invention to find the cost.

$100

$15

$25

$1

$1

$30

$30

$15

$10

$50

$20

$10

$5

$1

$10

Invention #10
This is a Slockomoggy

Look at the INVENTION on the Left.
It's for Sale.
How much does it cost?
$_____

Each of these parts below has a value.
Calculate the value of the items used
to create the invention to find the cost.

$100

$15

$25

$1

$1

$30

$30

$15

$10

$50

$20

$10

$5

$1

$10

CHAPTER THREE

THE GUESSING GAMES

Are you Naturally Brilliant?

ARE YOU GOOD at GUESSING?

Are you naturally brilliant or just good at calculating?
A true genius will be able to determine
the winner before calculating.

Both inventions have been entered into a competition.
You are the judge. They earn points for each special
part listed below. First you must **GUESS** which invention will be the winner. Then calculate the scores to
see if your guess was correct.

Number of Points for Each Part:

♫ = 3   ⬭ = 7   🌰 = 5

⬭ = 2   🍶 = 8   ◯ = 4

🍃 = 1   🪐 = 9   ☆ = 6

1

2

I think the win-
ner will be:
#____

I did the math
and the winner
is #____

ARE YOU GOOD at GUESSING?

Are you naturally brilliant or just good at calculating?
A true genius will be able to determine
the winner before calculating.

Both inventions have been entered into a competition.
You are the judge. They earn points for each special part
listed below. First you must **GUESS** which invention will
be the winner. Then calculate the scores to see if your
guess was correct.

Number of Points for Each Part:

♫ = 3   ⌣ = 7   🌰 = 5

⊖ = 2   🍾 = 8   ⬤ = 4

🍃 = 1   🪐 = 9   ☆ = 6

3

I think the win-
ner will be:
#____

I did the math
and the winner
is #____

4

ARE YOU GOOD at GUESSING?

Are you naturally brilliant or just good at calculating?
A true genius will be able to determine
the winner before calculating.

Both inventions have been entered into a competition.
You are the judge. They earn points for each special
part listed below. First you must **GUESS** which invention will be the winner. Then calculate the scores to
see if your guess was correct.

Number of Points for Each Part:

♫ = 3   ☕ = 7   🌰 = 5

⬭ = 2   🍶 = 8   🔵 = 4

🍃 = 1   🪐 = 9   ☆ = 6

5

I think the win-
ner will be:
#____

I did the math
and the winner
is #____

#2

6

ARE YOU GOOD at GUESSING?

Are you naturally brilliant or just good at calculating?
A true genius will be able to determine
the winner before calculating.

Both inventions have been entered into a competition.
You are the judge. They earn points for each special
part listed below. First you must **GUESS** which invention will be the winner. Then calculate the scores to
see if your guess was correct.

Number of Points for Each Part:

♫ = 3   ☕ = 7   🌰 = 5

▱ = 2   🍶 = 8   ◯ = 4

🍃 = 1   🪐 = 9   ☆ = 6

7

I think the winner will be:
#____

I did the math and the winner is
#____

8

ARE YOU GOOD at GUESSING?

Are you naturally brilliant or just good at calculating?
A true genius will be able to determine
the winner before calculating.

Both inventions have been entered into a competition.
You are the judge. They earn points for each special
part listed below. First you must **GUESS** which invention will be the winner. Then calculate the scores to
see if your guess was correct.

Number of Points for Each Part:

♫ = 3

☕ = 7

🌰 = 5

▭ = 2

🫙 = 8

⚪ = 4

🍃 = 1

🪐 = 9

☆ = 6

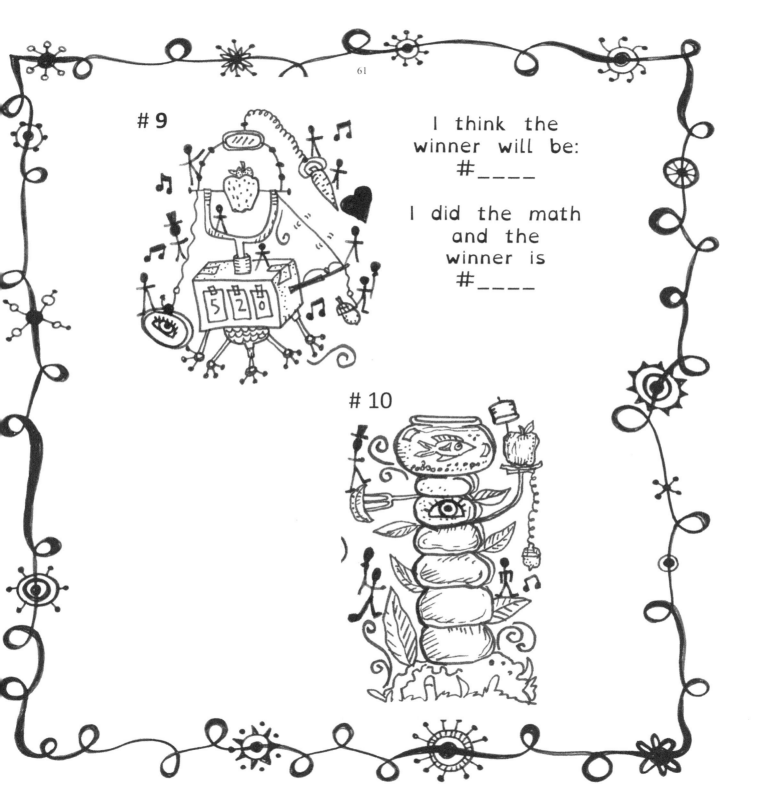

#9

I think the
winner will be:
#____

I did the math
and the
winner is
#____

10

Chapter FOUR

Mental Memory
challenge

HOW GOOD IS YOUR MEMORY?

Here are four Inventions that
you should be familiar with by now.

Can you remember the name of each
Invention without turning the page?
If you can't really remember you **must** make a guess,
at least you will have something to laugh about later.

1.
2.
3.
4.

What Invention cost the most to operate
in Chapter One?

Your Answer:_____

Correct Answer:_____

What Invention is the most expensive?

#_____

1

2

3

4

HOW GOOD IS YOUR MEMORY?

Here are four Inventions that
you should be familiar with by now.

Can you remember the name of each
Invention without turning the page?

If you can't really remember you **must** make a guess,
at least you will have something to laugh about later.

1.
2.
3.
4.

What Invention cost the most to operate
in Chapter One?

Your Answer:_____

Correct Answer:_____

What Invention is the most expensive?

_____

45

6

7

8

HOW GOOD IS YOUR MEMORY?

Here are four Inventions that
you should be familiar with by now.

Can you remember the name of each
Invention without turning the page?

If you can't really remember you **must** make a guess,
at least you will have something to laugh about later.

1.
2.
3.
4.

What Invention cost the most to operate
in Chapter One?

Your Answer:_____

Correct Answer:_____

What Invention is the most expensive?

_____

1

2

9

10

Chapter FIVE

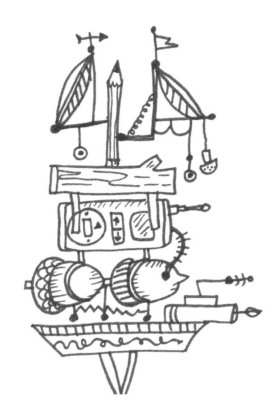

HIGH LEVEL
INTELLIGENCE INTEGRATION
CHALLENGE

Invention #11
This is a Fabiobuster

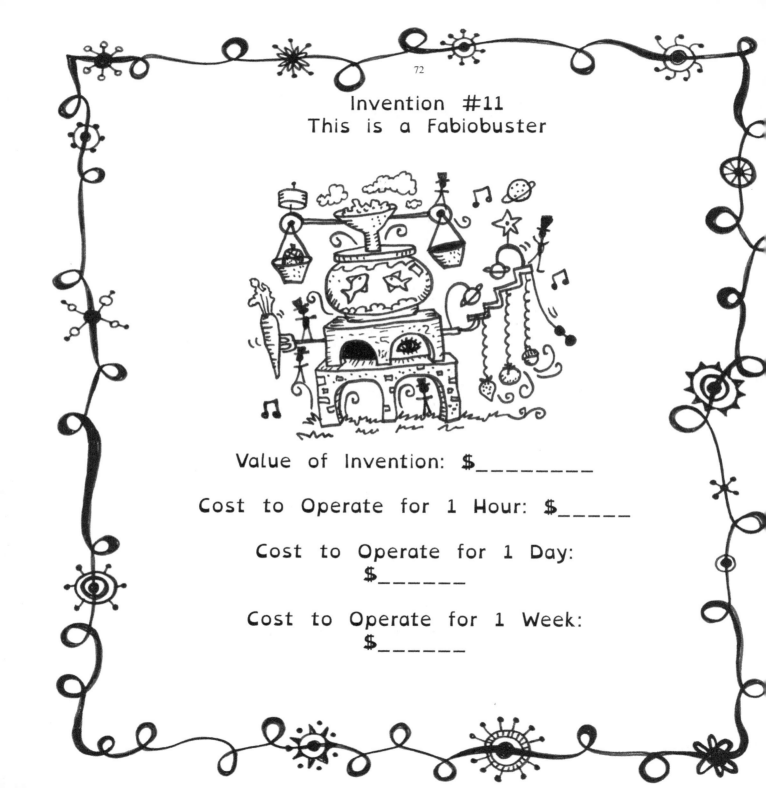

Value of Invention: $_____

Cost to Operate for 1 Hour: $_____

Cost to Operate for 1 Day:
$_____

Cost to Operate for 1 Week:
$_____

Value of Parts:

$100 $15 $25 $1 $1

$30 $30 $15 $10 $50

$20 $10 $5 $1 $10

Labor Cost:
You must pay each little employee
$10 per hour to operate this Invention.

Invention #11

FabioBuster FOR SALE!

Create an interesting advertisement
for this Invention on the next page.

Invention #12
This is a Stickolater

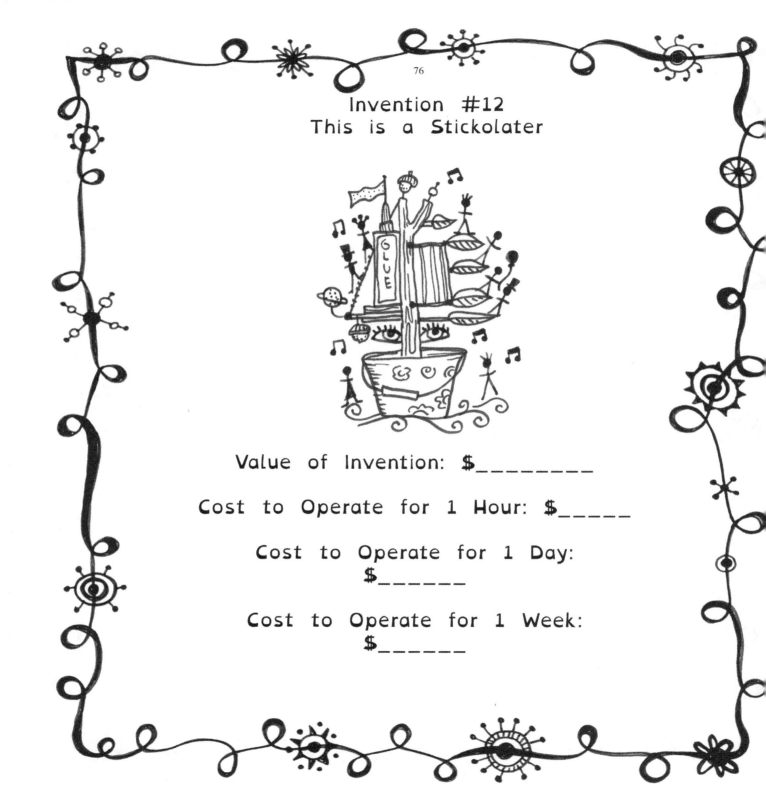

Value of Invention: $_____

Cost to Operate for 1 Hour: $_____

Cost to Operate for 1 Day:
$_____

Cost to Operate for 1 Week:
$_____

Value of Parts:

$100 $15 $25 $1 $1

$30 $30 $15 $10 $50

$20 $10 $5 $1 $10

Labor Cost:
You must pay each little employee
$5.00 per hour to operate this Invention.

Invention #12

Stickolater

FOR SALE!

Create an interesting advertisement
for this Invention on the next page.

79

Invention #13
This is a Whistlewampus

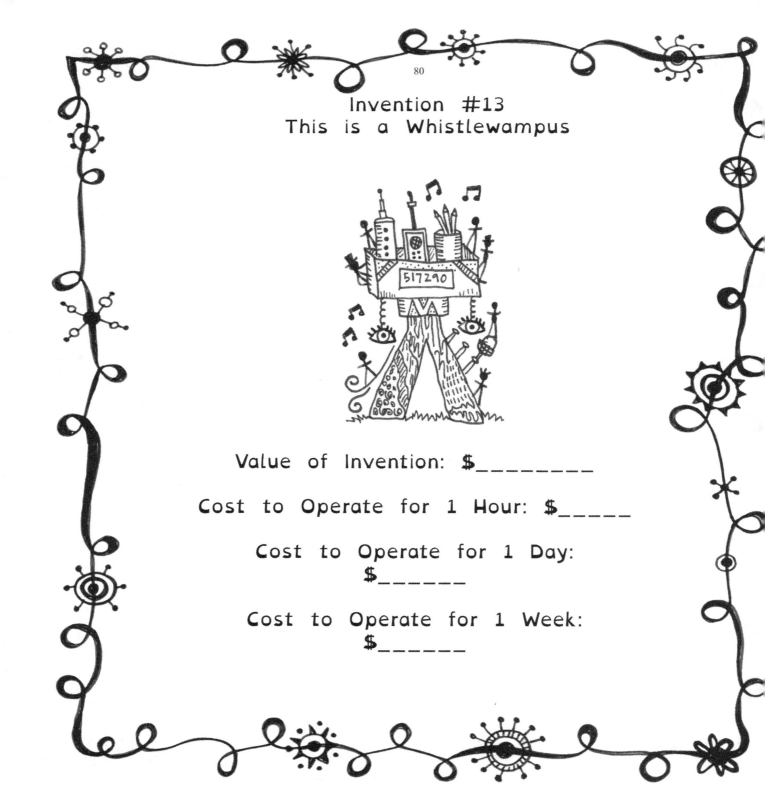

Value of Invention: $_____

Cost to Operate for 1 Hour: $_____

Cost to Operate for 1 Day:
$_____

Cost to Operate for 1 Week:
$_____

Value of Parts:

$100 $15 $25 $1 $1

$30 $30 $15 $10 $50

$20 $10 $5 $1 $10

Labor Cost:
You must pay each little employee
$11.99 per hour to operate this Invention.

Invention #13

WHISTLEWAMPUS

FOR SALE!

Create an interesting advertisement
for this Invention on the next page.

83

Invention #14
This is a Dapper-splatter

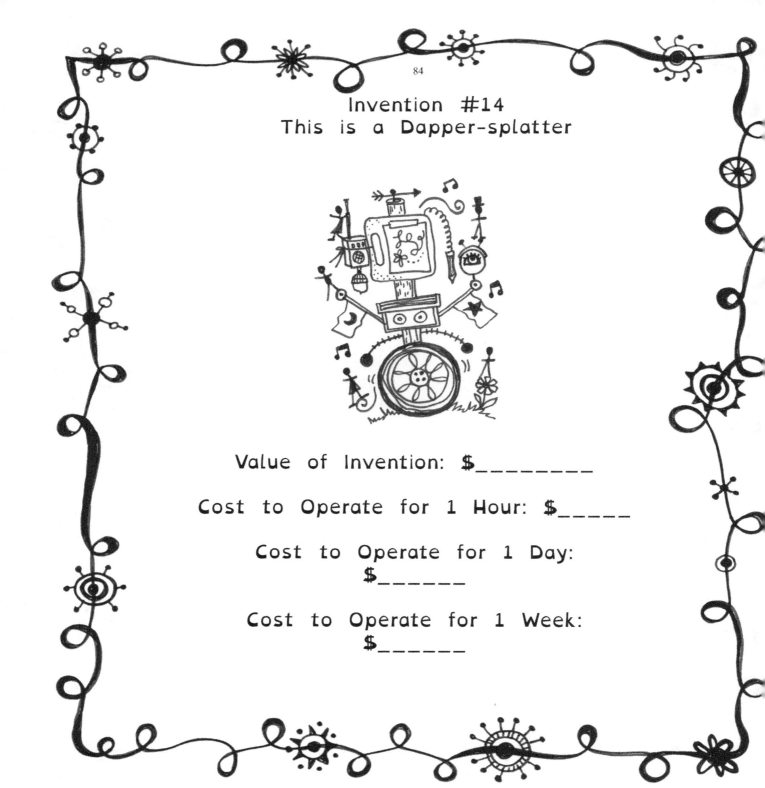

Value of Invention: $_____

Cost to Operate for 1 Hour: $_____

Cost to Operate for 1 Day:
$_____

Cost to Operate for 1 Week:
$_____

Value of Parts:

$100

$15

$25

$1

$1

$30

$30

$15

$10

$50

$20

$10

$5

$1

$10

Labor Cost:
You must pay each little employee
$3.75 per hour to operate this Invention.

Invention #14

Dapper-splatter
for SALE!

Create an interesting advertisement
for this Invention on the next page.

87

Invention #15
This is a Rethinkerblinger

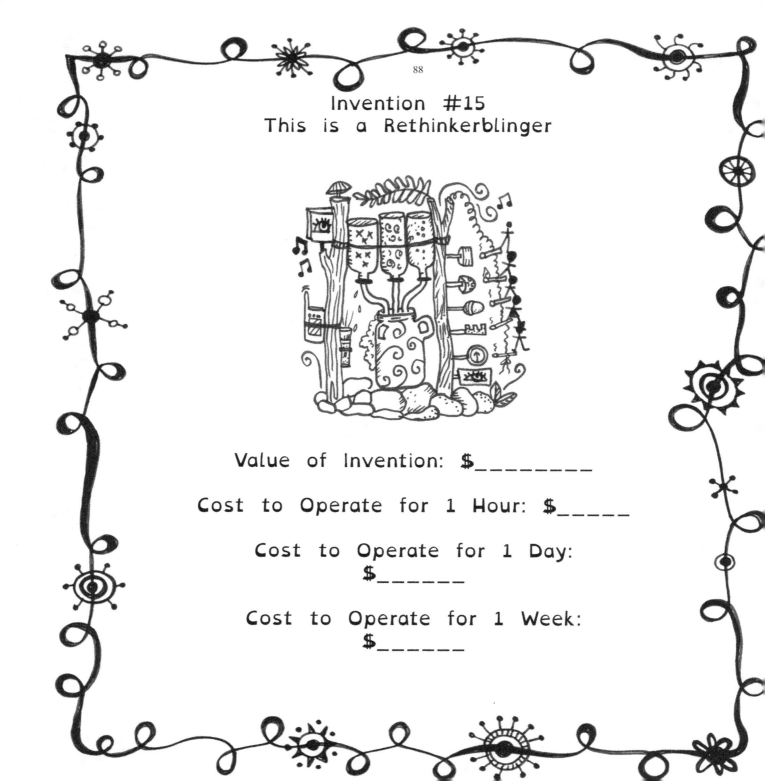

Value of Invention: $_____

Cost to Operate for 1 Hour: $_____

Cost to Operate for 1 Day:
$_____

Cost to Operate for 1 Week:
$_____

Value of Parts:

Labor Cost:
You must pay each little employee
$2.00 per hour to operate this Invention.

Invention #15

RETHINKERBLINGER

FOR SALE!

Create an interesting advertisement
for this Invention on the next page.

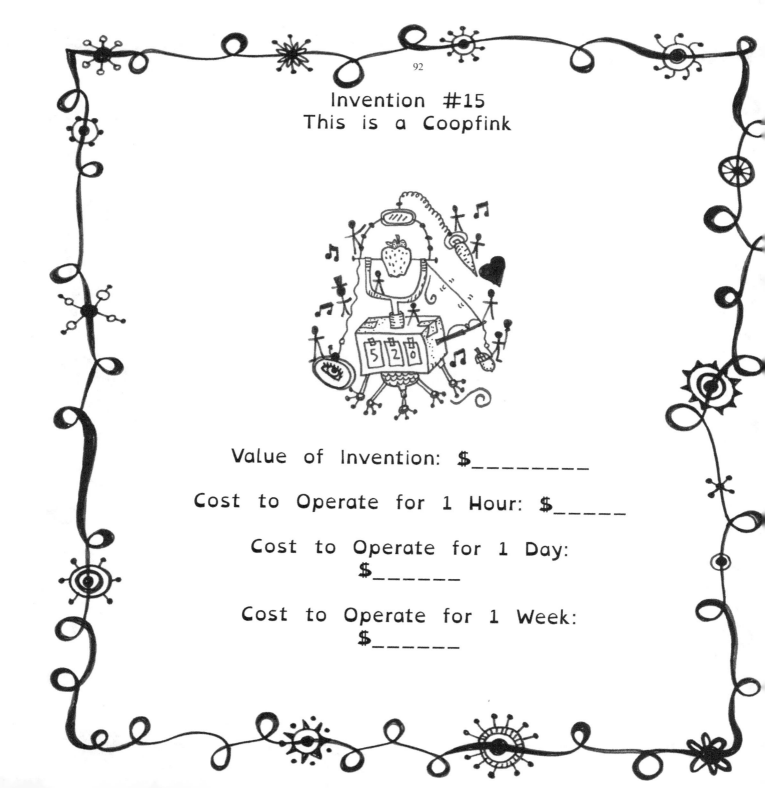

Invention #15
This is a Coopfink

Value of Invention: $_____

Cost to Operate for 1 Hour: $_____

Cost to Operate for 1 Day:
$_____

Cost to Operate for 1 Week:
$_____

Value of Parts:

$100

$15

$25

$1

$1

$30

$30

$15

$10

$50

$20

$10

$5

$1

$10

Labor Cost:
You must pay each little employee
$7.00 per hour to operate this Invention.

Invention #16

COOPFINK
FOR SALE!

Create an interesting advertisement
for this Invention on the next page.

95

Invention #16
This is a Sirconibill

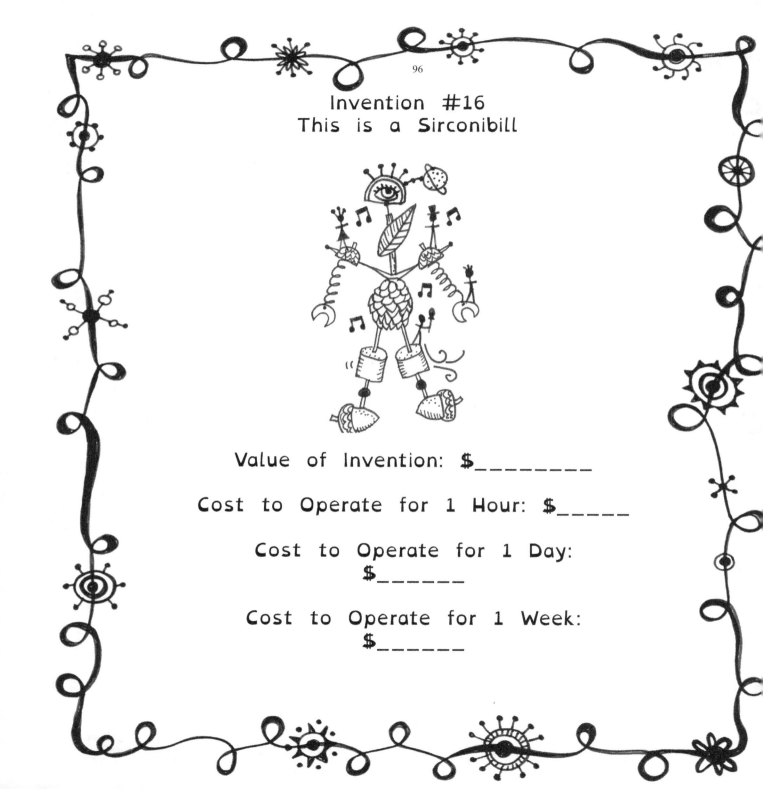

Value of Invention: $_____

Cost to Operate for 1 Hour: $_____

Cost to Operate for 1 Day:
$_____

Cost to Operate for 1 Week:
$_____

Value of Parts:

$100

$15

$25

$1

$1

$30

$30

$15

$10

$50

$20

$10

$5

$1

$10

Labor Cost:
You must pay each little employee
$4.00 per hour to operate this Invention.

Invention #17

SIRCONIBILL

FOR SALE!

Create an interesting advertisement
for this Invention on the next page.

99

CHAPTER SIX

HIGH LEVEL
NARRATIVE COMMUNICATION
CHALLENGE

Invention #17

TALE OF THE
DIGGIDY-TOOPLE

Write a short story about
this Invention on the next page.

103

Invention #18

TaLE OF tHE
CORGOBaCKER

Write a short story about
this Invention on the next page.

105

Invention #19

Tale of the Gootiefab

Write a short story about
this Invention on the next page.

107

Invention #20

Tale of the

Coblozzybark

Write a short story about
this Invention on the next page.

109

Chapter Seven

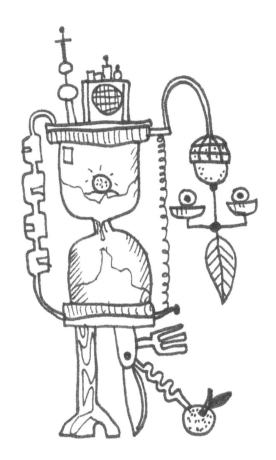

Projective
Visual thinking
Challenge

Invention #21

ROtitOOtBOZZER

Use a black gel pen to
DRAW the missing parts.

Invention #22

Caulipoppins

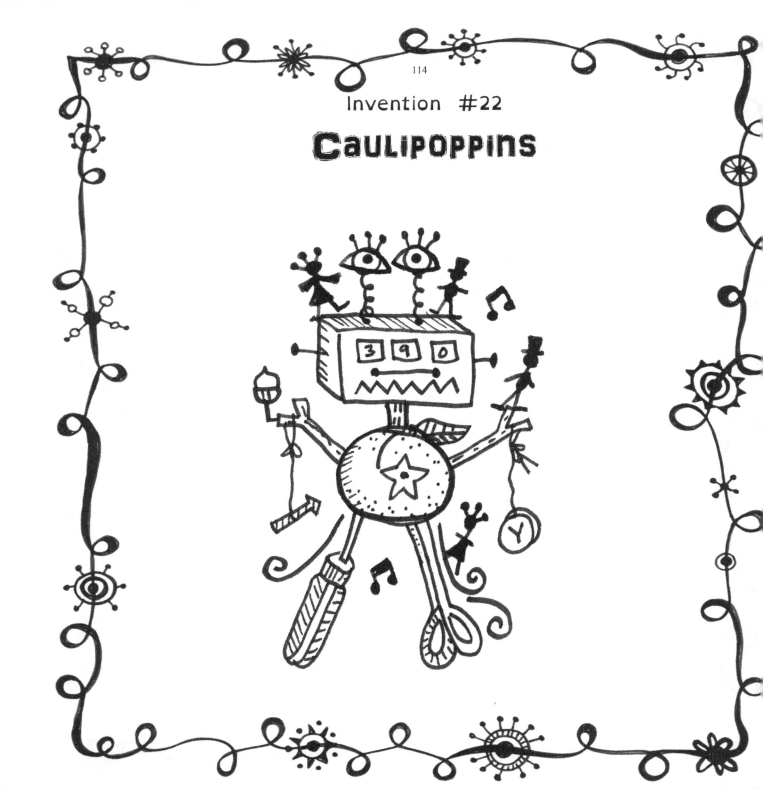

Use a black gel pen to
DRAW the missing parts.

Invention #23

LERptESteR

Use a black gel pen to
DRAW the missing parts.

Invention #24

WHAM-BAMSPATTER

Use a black gel pen to
DRAW the missing parts.

Invention #25

KUMBUYAH

Use a black gel pen to
DRAW the missing parts.

CHaPtER EIGHt

ARE YOU a GaME CHaNGER?

INVEntiVE ABILitiES EVaLuation
& CREatiVE APPLICation CHaLLENGE

HOW CREATIVE ARE YOU?

The highest form of intelligence
Lies within one's ability to create.

So, can you Invent a game to play alone or with a friend? Use the pictures and information on each of the following pages to create a game.

You can use a black gel pen to add more elements to the game. You can use dice, or anything else you can imagine.

How many different kinds of games
can you invent?

👁 = 100

🧍 = 50

🎩 = 40

🙋 = 30

🧍 = 20

🧍 = 10

🍃 = 1

⬤ = 2

♫ = 3

◯ = 4

🌰 = 5

☆ = 6

☕ = 7

🍼 = 8

🪐 = 9

 = 100

= 50

= 40

= 30

= 20

= 10

 = 1

= 2

♫ = 3

= 4

= 5

☆ = 6

= 7

= 8

= 9

👁 = 100

🧍 = 50

🎩 = 40

🙆 = 30

🧍 = 20

🧍 = 10

🍃 = 1

⬭ = 2

♫ = 3

◓ = 4

🌰 = 5

☆ = 6

☕ = 7

🍶 = 8

🪐 = 9

$=100$

$=50$

$=40$

$=30$

$=20$

$=10$

$=1$

$=2$

$=3$

$=4$

$=5$

$=6$

$=7$

$=8$

$=9$

👁 = 100

🧍 = 50

🎩 = 40

🙆 = 30

🧍 = 20

🧍 = 10

🍃 = 1

⬭ = 2

♫ = 3

◯ = 4

🌰 = 5

☆ = 6

☕ = 7

🍶 = 8

🪐 = 9

$\text{eye} = 100$

$\text{person (arms up)} = 50$

$\text{person (top hat)} = 40$

$\text{person (balloon)} = 30$

$\text{person} = 20$

$\text{person (simple)} = 10$

$\text{leaf} = 1$

$\text{container} = 2$

$\text{note} = 3$

$\text{ball} = 4$

$\text{acorn} = 5$

$\text{star} = 6$

$\text{cup} = 7$

$\text{bottle} = 8$

$\text{planet} = 9$

👁 = 100

👤 = 50

👤 = 40

👥 = 30

👤 = 20

👤 = 10

🍃 = 1

◎ = 2

♫ = 3

◯ = 4

🌰 = 5

☆ = 6

☕ = 7

🍶 = 8

🪐 = 9

= 100

= 50

= 40

= 30

= 20

= 10

= 1

= 2

= 3

= 4

= 5

= 6

= 7

= 8

= 9

=100

=50

=40

=30

=20

=10

=1

=2

♫ =3

=4

=5

☆=6

=7

=8

∅=9

$$\text{(eye)} = 100$$

$$\text{(figure)} = 50$$

$$\text{(figure)} = 40$$

$$\text{(figure)} = 30$$

$$\text{(figure)} = 20$$

$$\text{(figure)} = 10$$

$$\text{(leaf)} = 1$$

$$\text{(container)} = 2$$

$$\text{(music note)} = 3$$

$$\text{(ball)} = 4$$

$$\text{(acorn)} = 5$$

$$\text{(star)} = 6$$

$$\text{(cup)} = 7$$

$$\text{(bottle)} = 8$$

$$\text{(planet)} = 9$$

👁 = 100

🧍 = 50

🎩 = 40

🧍 = 30

🧍 = 20

🧍 = 10

🍃 = 1

⬭ = 2

♫ = 3

◯ = 4

🌰 = 5

☆ = 6

☕ = 7

🍼 = 8

🪐 = 9

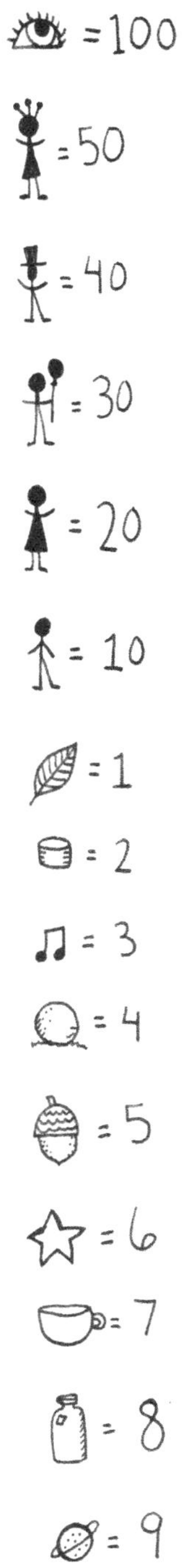

= 100

= 50

= 40

= 30

= 20

= 10

= 1

= 2

= 3

= 4

= 5

= 6

= 7

= 8

= 9

👁 = 100

🧍 = 50

🎩 = 40

🧍 = 30

🧍 = 20

🧍 = 10

🍃 = 1

⬭ = 2

♫ = 3

◯ = 4

🌰 = 5

☆ = 6

☕ = 7

🍼 = 8

🪐 = 9

= 100

= 50

= 40

= 30

= 20

= 10

= 1

= 2

♫ = 3

= 4

= 5

☆ = 6

= 7

= 8

= 9

👁 = 100

🧍 = 50

🎩 = 40

🧍 = 30

🧍 = 20

🧍 = 10

🍃 = 1

⬭ = 2

♫ = 3

◯ = 4

🌰 = 5

☆ = 6

☕ = 7

🍼 = 8

🪐 = 9

$\text{👁} = 100$

$\text{🧍} = 50$

$\text{🧍} = 40$

$\text{🧍} = 30$

$\text{🧍} = 20$

$\text{🧍} = 10$

$\text{🍃} = 1$

$\text{⬭} = 2$

$\text{♫} = 3$

$\text{◯} = 4$

$\text{🌰} = 5$

$\text{☆} = 6$

$\text{☕} = 7$

$\text{🍼} = 8$

$\text{🪐} = 9$

= 100

= 50

= 40

= 30

= 20

= 10

= 1

= 2

♫ = 3

= 4

= 5

☆ = 6

= 7

= 8

= 9

=100

= 50

= 40

= 30

= 20

= 10

= 1

= 2

♩ = 3

= 4

= 5

☆ = 6

= 7

= 8

= 9

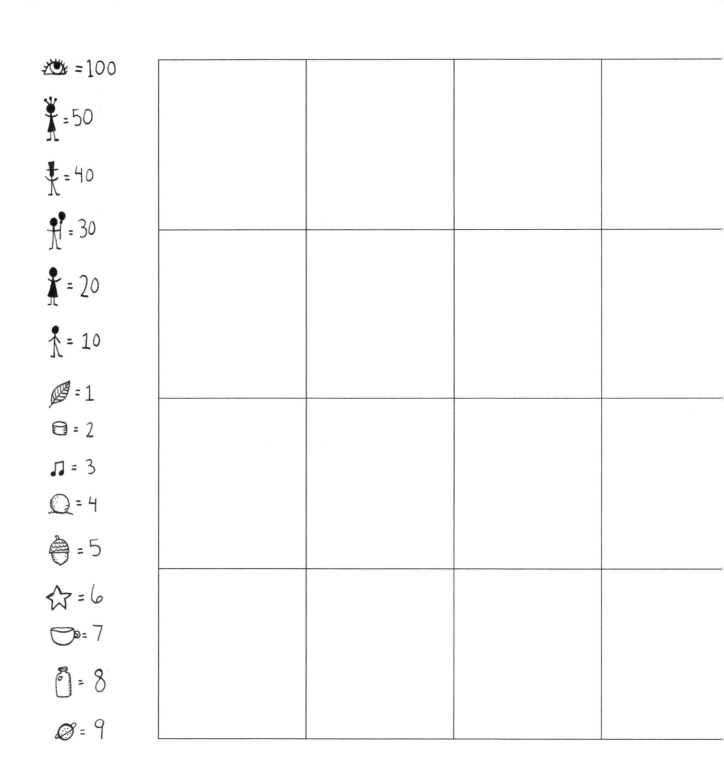

=100

=50

=40

=30

=20

= 10

=1

= 2

= 3

= 4

= 5

= 6

= 7

= 8

= 9

👁 = 100

🧍 = 50

🧍 = 40

🧍 = 30

🧍 = 20

🧍 = 10

🍃 = 1

🪣 = 2

♫ = 3

⚪ = 4

🌰 = 5

☆ = 6

☕ = 7

🍶 = 8

🪐 = 9

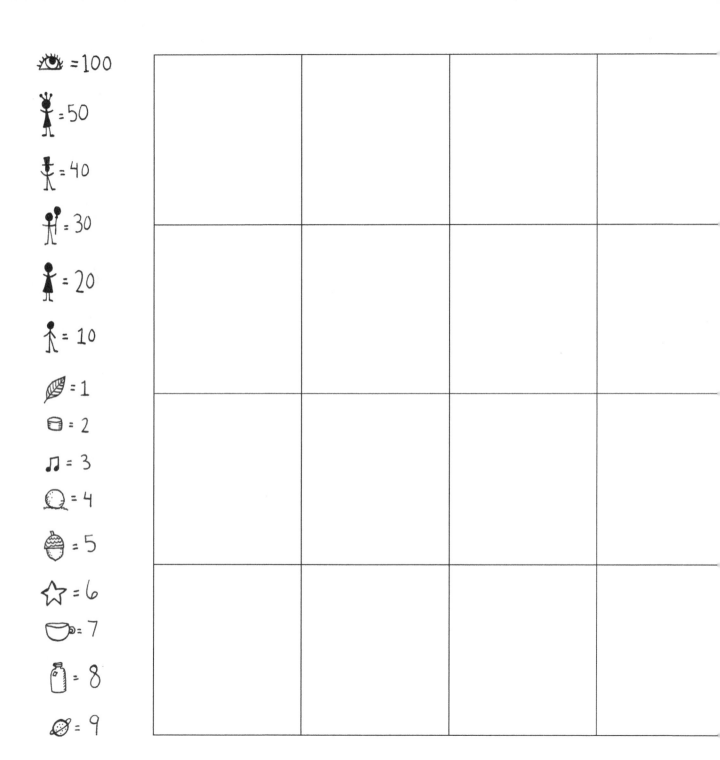

👁 = 100

🧍 = 50

🧍 = 40

🧍 = 30

🧍 = 20

🧍 = 10

🍃 = 1

⬭ = 2

♫ = 3

◯ = 4

🌰 = 5

☆ = 6

☕ = 7

🍼 = 8

🪐 = 9

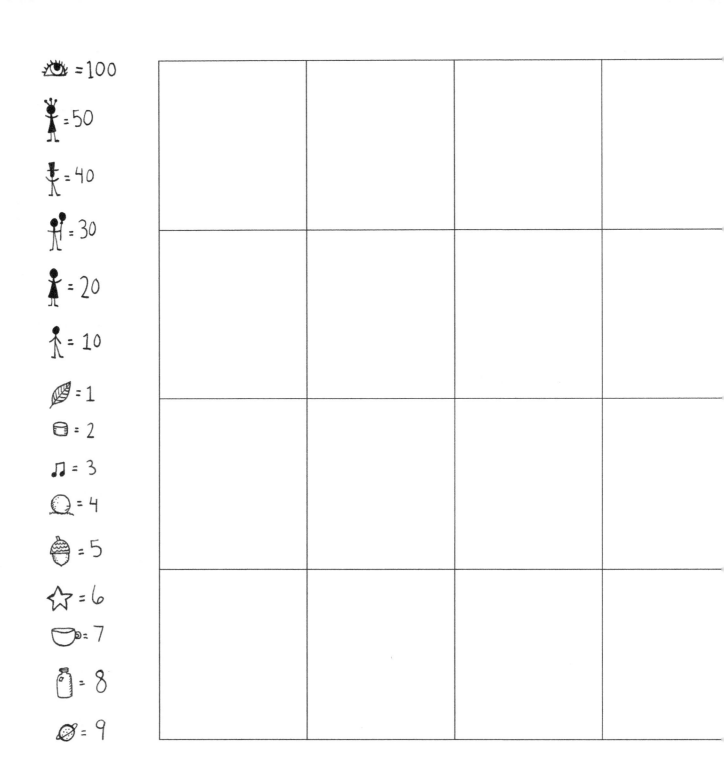

= 100

 = 50

= 40

= 30

= 20

= 10

= 1

= 2

♫ = 3

= 4

= 5

☆ = 6

= 7

= 8

= 9

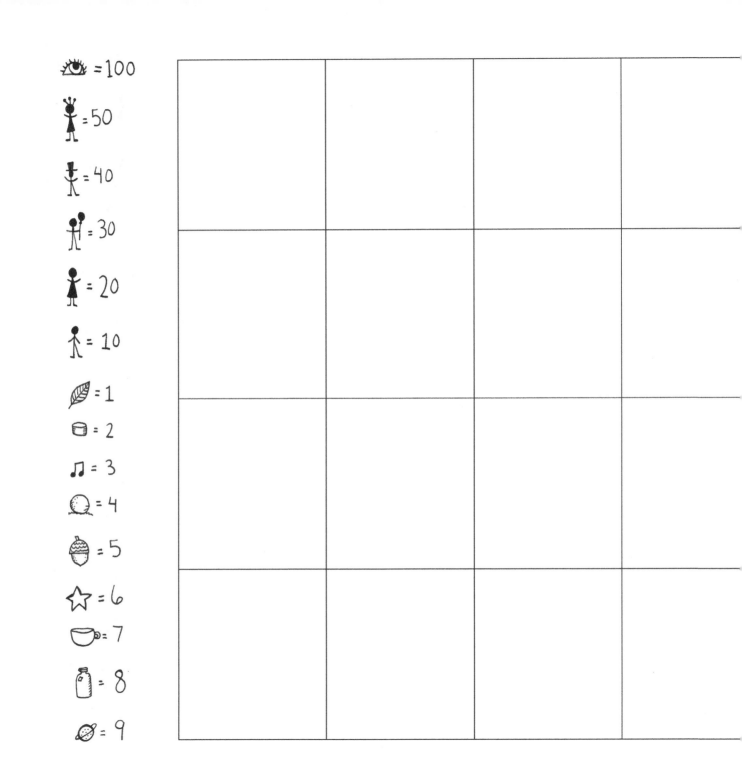

👁 = 100

🧍 = 50

🧍 = 40

🧍 = 30

🧍 = 20

🧍 = 10

🍃 = 1

⬭ = 2

♫ = 3

◯ = 4

🌰 = 5

☆ = 6

☕ = 7

🍼 = 8

🪐 = 9

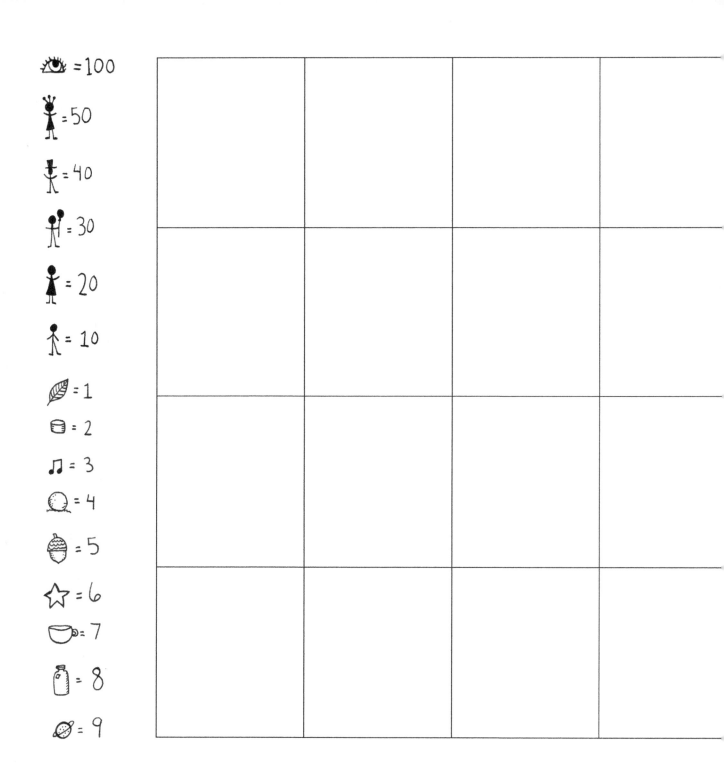

👁 = 100

🧍 = 50

🧍 = 40

🧍 = 30

🧍 = 20

🧍 = 10

🍃 = 1

⬭ = 2

♫ = 3

◯ = 4

🌰 = 5

☆ = 6

☕ = 7

🍼 = 8

🪐 = 9

154

PAPER FOR YOUR CALCULATIONS AND FOR DESIGNING YOUR OWN INVENTIONS.

This graph paper helps you to keep
your numbers well organized.
You may also use this
paper for calculations you
can design something new,
like an invention of your own.

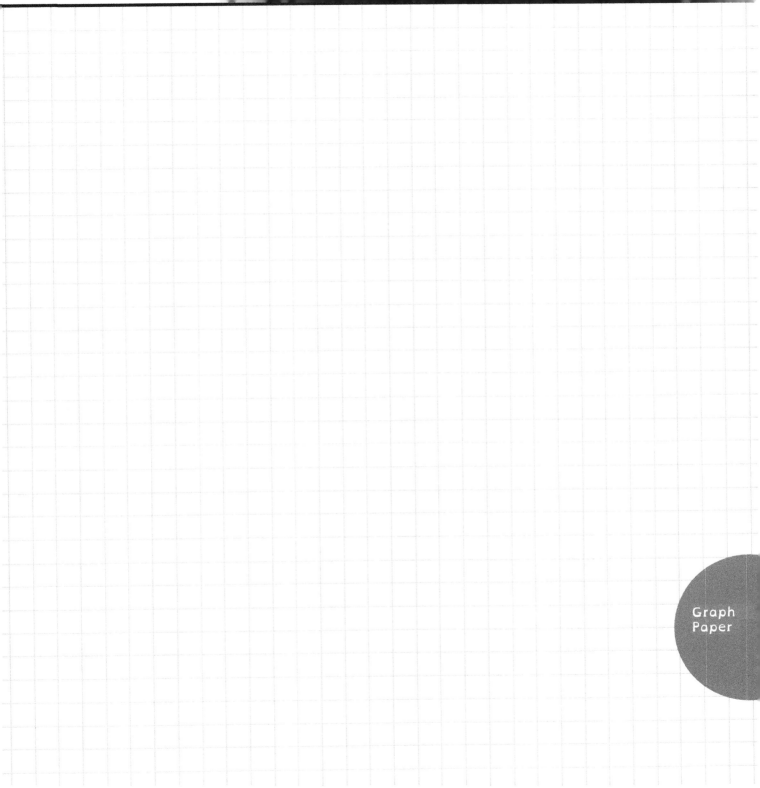

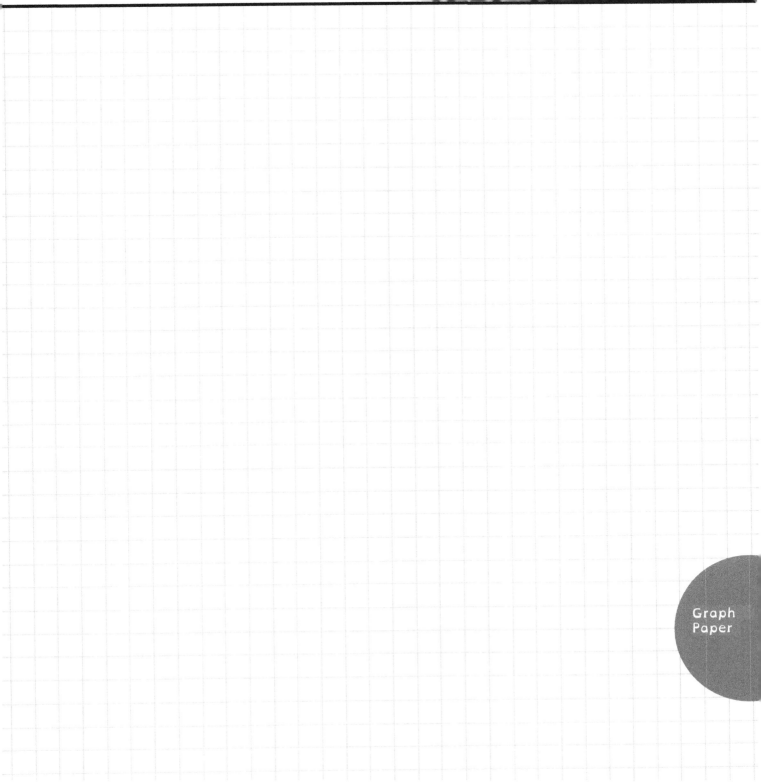

Graph
Paper

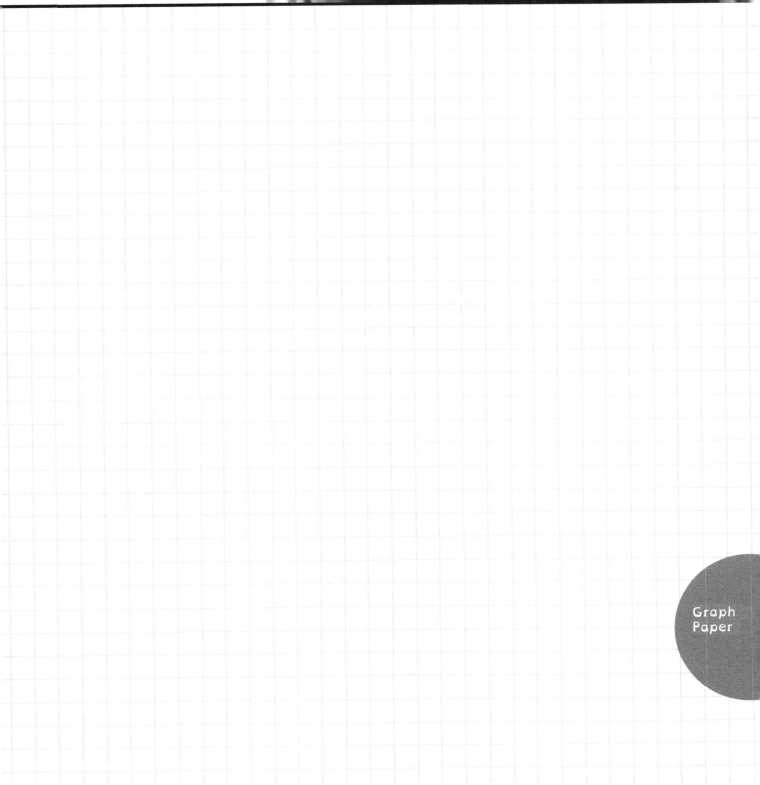

Graph
Paper

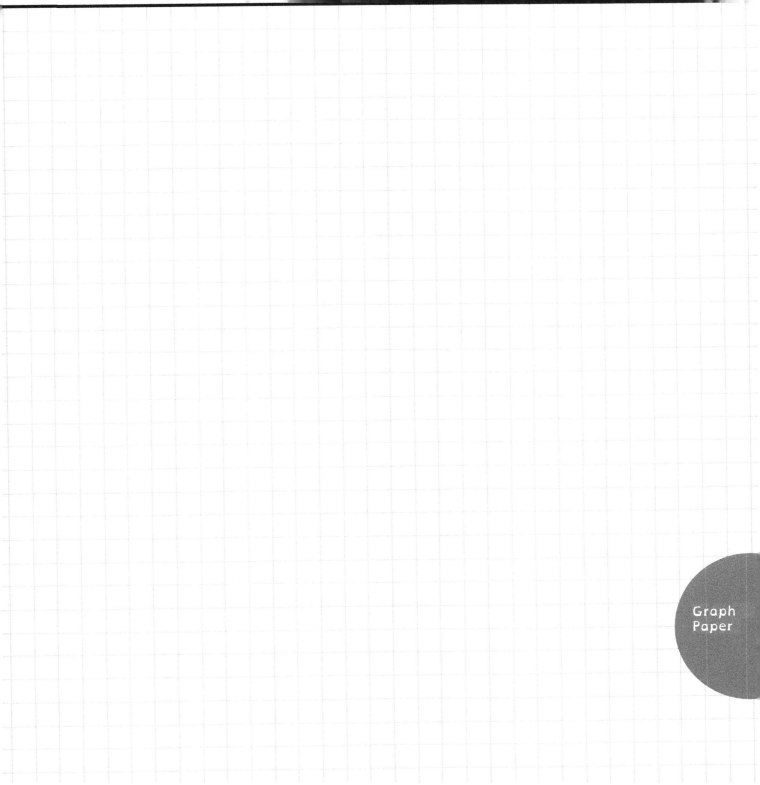

Graph
Paper

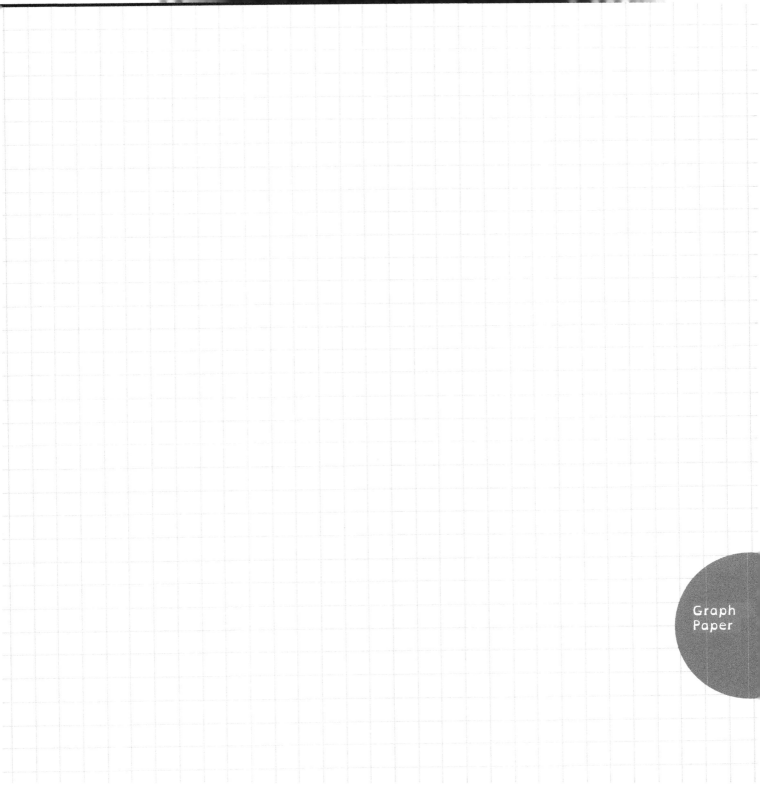

Graph
Paper

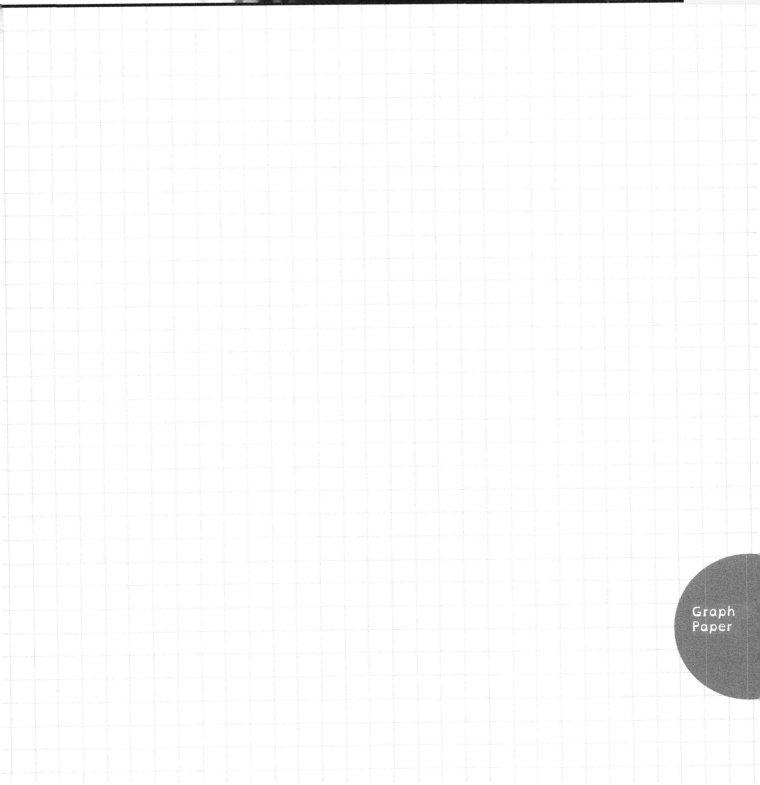

Graph
Paper

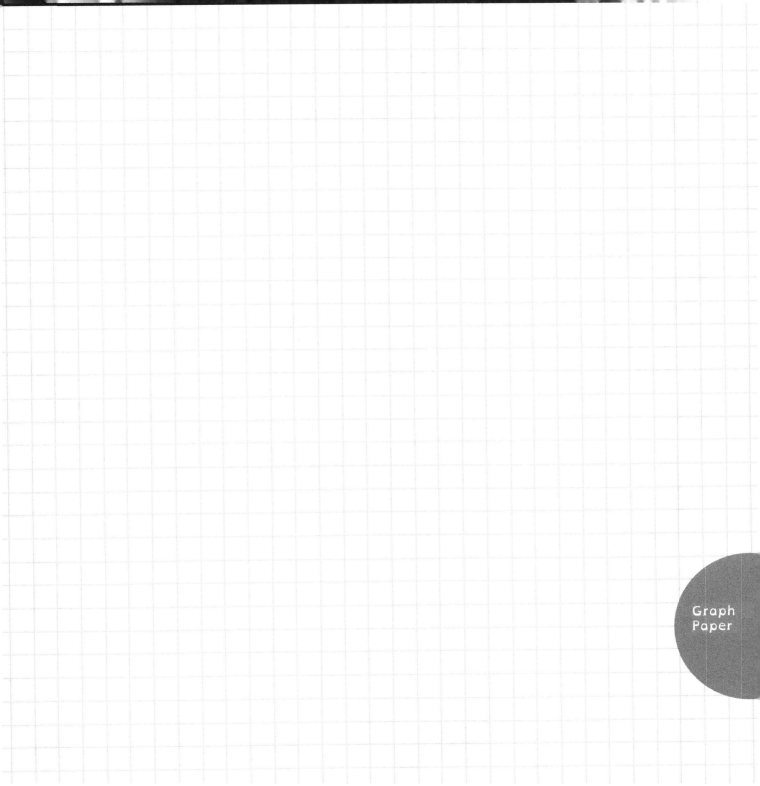

Chapter One- ANSWER KEY

Invention #1 = $105

Invention #2 = $240

Invention #3 = $504

Invention #4 = $1080

Invention #5 = $1,200

Invention #6 = $175

Invention #7 = $5,250

Invention # 8 = $168,000

Invention #9 = $146.25

Invention #10 = $58,500

Use a Calculator to check all your other answers.

Sorry, real life has no answer key, no multiple choice, no grades, and no cheats!

Welcome to real life. How did you do? There are no answers to the bonus questions. They are relative and depend on logic and creative thinking. Ask a family member, friend or the guy beside you on the plane to review your answers. No friends to help? Practice your self-evaluation skills!

188

The Thinking Tree

Made in the USA
Las Vegas, NV
28 October 2024